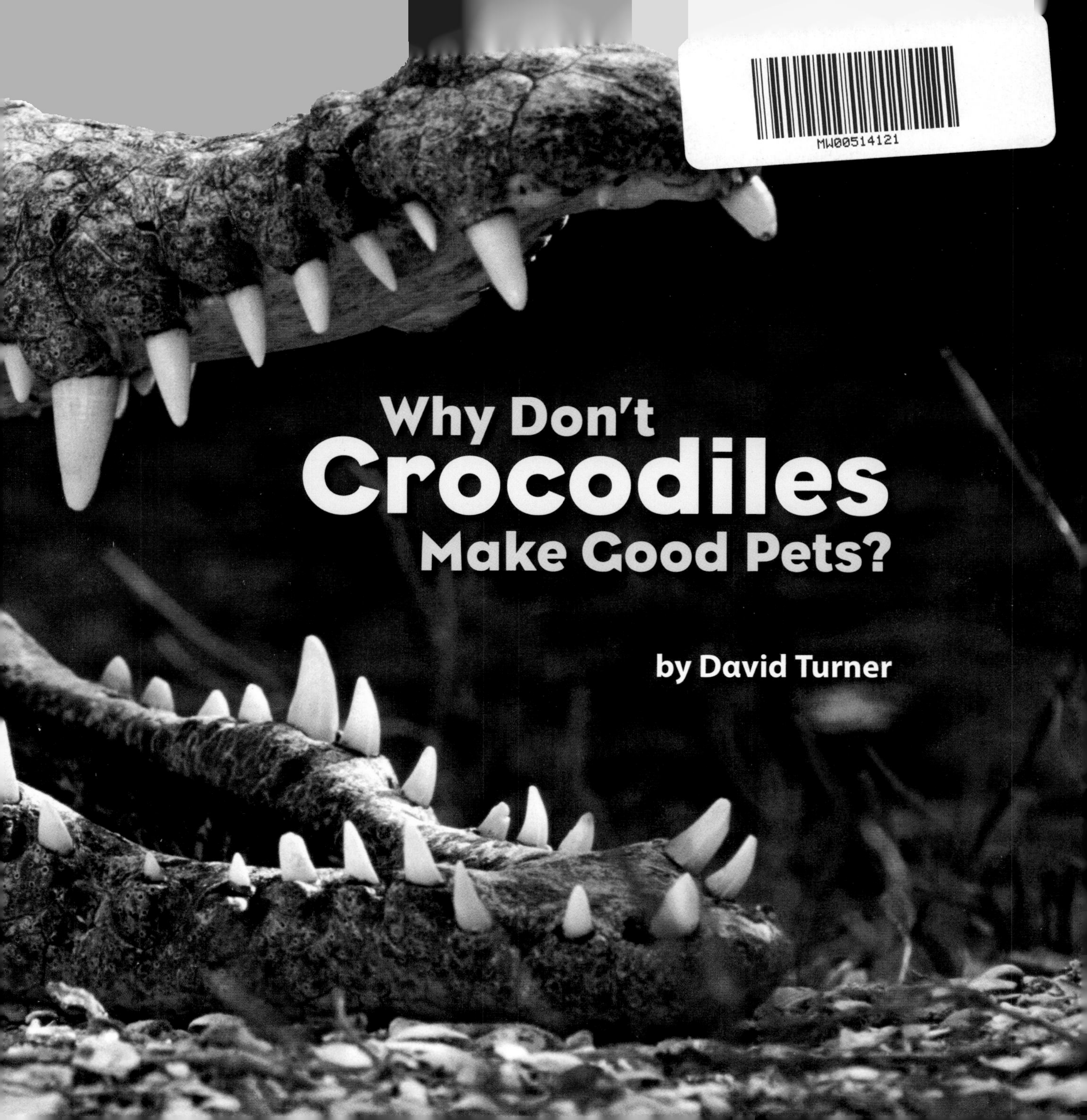

Why Don't Crocodiles Make Good Pets?

by David Turner

Baby animals are cute.

Even baby crocodiles!

But is one cute enough to hold in the palm of your hand?

Not for long!

You might be able to hold a baby crocodile, or hatchling, in the palm of your hand at first. But an adult crocodile would grow to be too big.

Would it be difficult for an adult crocodile to find something to eat?

Not at all.

A grown-up crocodile eats just about anything.

So why don't crocodiles make good pets?

Because they're crocodiles!

From Egg to Adult

- A crocodile hatches from an egg.
- A baby crocodile can stay in its nest for eight weeks.
- A crocodile's life cycle can last as long as 70 years.